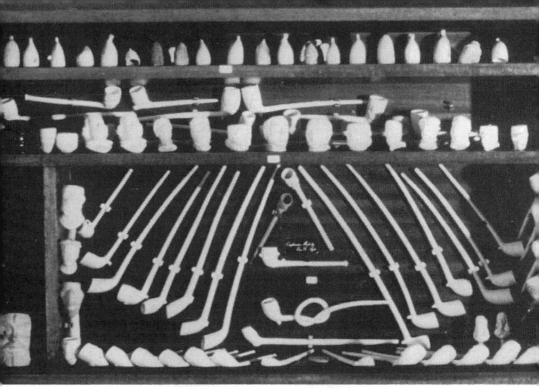

Pipes displayed in a wall cabinet from Maurice Moverley's collection. At bottom left note the large bowl depicting the head of Father Christmas. The knotted stem pipe at centre was made specially for this collection by the author.

T0174458

CLAY TOBACCO PIPES

Eric G. Ayto

Shire Publications Ltd

CONTENTS

Published by Shire Publications Ltd, Midland House, West Way, Botley, Oxford 0X2 0PH, Website: www.shirebooks.co.uk
Copyright © 1979 and 1994 by Eric G. Ayto. First edition 1979; reprinted 1984. Second edition 1987; reprinted 1990. Third edition 1994; reprinted 1999 and 2002. Transferred to digital print on demand 2011. Shire Library 37. ISBN 978 0 74780 248 8.

Printed and bound in Great Britain.

British Library Cataloguing in Publication Data: Ayto, Eric G. Clay tobacco pipes. – 3rd ed. – (Shire album; 37) 1. Clay tobacco-pipes–Great Britain–History. I. Title 688'.42 TS2270. ISBN-10 0 7478 0248 3. ISBN-13 978 0 74780 248 8.

ACKNOWLEDGEMENTS

The author wishes to thank the staff of Gosport Museum, for their kind assistance and for the loan of locally made pipes illustrated in this book, and Maurice Moverley, for the use of pipes from his private collection. Photographs are acknowledged as follows: Tony Bird, page 1; Mrs I. Webb, page 13; Winchester City Museum, page 19; Keeling and Freemantle, pages 21 and 22; Piet J. G. Tengnagel, CPCC-International, page 26; Peter Hammond, SCPR, pages 27 and 28.

MIX
Paper from responsible sources
FSC FSC® C013604
www.fsc.org

Cover: Clay tobacco pipes were often produced to commemorate famous personalities of the day. The three examples shown on the cover depict (from top to bottom): King Edward VII, Queen Alexandra and General Gordon, made famous in the siege of Khartoum, where he died in 1885. (Photograph courtesy of David Higgins)

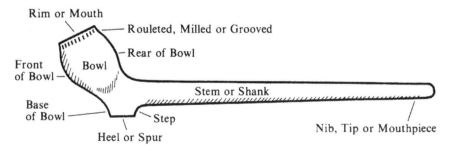

The different parts of a clay tobacco pipe. An early seventeenth-century style is illustrated.

English Civil War Society members smoking clay pipes during a muster at Powderham Castle, Devon.

INTRODUCTION

For a number of years now there has been a growing interest in the collection of everyday objects of the past as well as in their history and origin. Not least among these is the clay tobacco pipe, which has, from its humble beginnings to its more sophisticated forms in the nineteenth century, played a prominent part in social life for over three hundred years.

Since this book was first published in 1979 a great deal of detailed and painstaking research has been carried out by members of the Society for Clay Pipe Research and other historical organisations, bringing to light fresh information and evidence on known pipemakers as well as finding new ones. There has also been a noticeable increase in the interest in and collection of clay pipes from other parts of Europe, and it is largely for this reason that this book has been reissued to include a section on pipes made in Holland, Belgium, Germany and France, where, as in England, the huge demand warranted pipemaking industries of their own.

The history of clay tobacco pipes and the people who made them is a fascinating subject and in relating their history and evolution, and providing a background to the craft and to the pipemakers, this book has proved to be of considerable interest and value to the casual and serious collector alike and it is hoped that it will long continue to do so.

ORIGIN AND DEVELOPMENT

No one knows who made the first clay pipe for smoking tobacco, but the idea was probably adopted from the American Indians about the middle of the sixteenth century. There is little doubt that the craft of making clay tobacco pipes began in England shortly after the introduction of tobacco (about 1558) in order to satisfy the demand of people, including women and children, to take up the art and pleasure of 'tobacco drinking', as it was then called.

Although the principal form of the clay pipe remained much the same throughout its long life, notable variations in the style and size of the bowls occurred, as well as variations in the length of the stems. Some styles were the consequence of changing fashions, but others could well have been the result of improved skills of both the pipemaker and the mouldmaker. The size and capacity of the bowl itself, however, was influenced by the cost and availability of tobacco at the time.

The earliest description of the English clay pipe was in 1573 by William Harrison in his *Great Chronologie*. Here he describes the pipe as being 'an Instrument formed like a ladell'. This spoon-like shape was probably derived from the Indian pipe, which was used only for medicinal purposes and religious ceremonies.

By 1580 the bowl had altered to provide a better container for the tobacco, adopting a rather ingenious barrel shape and a forward incline. The base of the bowl was flat and the stem was straight and only about 4 to 6 inches (100 to 150mm) long; the pipe could therefore rest upright on the table. Although these early pipes were very small and rather crude, they were nonetheless extremely functional. The inside diameter of the bowl was about $\frac{1}{4}$ inch (6mm) and the bore of the stem around $\frac{1}{8}$ inch (3 mm).

Several names have been given to these tiny pipes according to the locality in which they were found and the beliefs of the finders: fairy pipes, elfin pipes, old man's pipes, Celtic pipes, Cromwellian pipes and even Roman pipes. They are sometimes referred to as *plague pipes* because of the large numbers found in plague pits during excavation work in London: people were encouraged to smoke clay pipes in those days in the belief that it would ward off the disease.

By 1640 the inside diameter of the bowl had increased only to about $\frac{3}{8}$ inch (9 mm) and there was no noticeable increase in the length of the stem. After this date the bowl became much larger and the stem longer (about 10 to 14 inches, 250 to 350 mm) and, except for the development of a short rounded spur in place of the flat heel and a few minor variations in style in different parts of England, the basic shape remained the same for the next sixty years or so. The reason for the spur is obscure; perhaps, because of the longer stem, the bowl was allowed to rest on the table when being smoked, thus preventing the heat from the bowl spoiling the polished surface.

A milled or plain ring at the top of the bowl was common for this period and, except for the occasional maker's mark, the majority of seventeenth-century pipes were otherwise plain. Towards the end of the century the inside diameter of the bowl was about $\frac{1}{2}$ inch (13 mm) and the bore diameter of the stem from about $\frac{3}{32}$ to $\frac{1}{8}$ inch (2.4 to 3 mm). Most stems were straight, but some tended to curve either upwards or downwards, more probably from distortion during firing than from design, and the bowl had lost its bulbous look in favour of a more elongated appearance.

A few elaborately decorated pipes were made during the first half of the century and, although there were some English versions, they are thought to be mainly of Dutch origin. The designs were either stamped or incised by hand (on both bowl and stem) or moulded in relief. Two well-known examples are the head of Jonah about to be swallowed by a crocodile or serpent (which might well represent King James I, who tried hard to suppress the habit of smoking tobacco) and a rather charming pipe showing what might be the faces of Charles I and Henrietta Maria and which may have been issued to

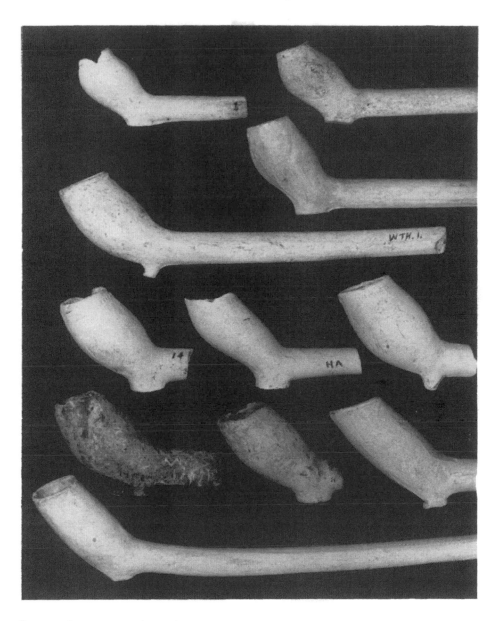

Seventeenth-century pipes (top to bottom and left to right): c. 1620-40, small bowl with flat base; c. 1640-60, slightly larger bowl and stem with flat heel; c. 1660-80, larger capacity bowl with flat heel; c. 1660-80, large capacity bowl with forward protruding spur; c. 1660-80, similar size and shape of bowl but with flat heel; c. 1660-80, west country type with flared heel; c. 1660-80, west country or midlands type with rounded spur; c. 1660-80, robust bowl and stem, similar to pipes found in the midlands, with rounded spur; c. 1680-90, development of parallel bowl with flat heel; c. 1680-1700, long parallel bowl with degenerate pedestal spur; c. 1680-1700, long parallel bowl with protruding heel.

commemorate their wedding in 1625.

Shortly after 1700 some important changes in quality took place. Pipes were being made with more accurate dimensions, a smoother finish and a higher degree of brittleness. The wall of the bowl was thinner and the stem more slender. All this suggests a steady improvement in the skills of the craft, including firing techniques, as well as in the art of the mouldmaker. At about the same time the top of the bowl was being trimmed level with the stem instead of sloping at a downward angle towards the front of the bowl. This particular change, together with a more upright bowl than before, is an indication of a new method of manufacture which was probably brought about by the introduction of the gin-press, as described later.

Early eighteenth-century pipes favoured a flat-bottomed (pedestal) spur and some were produced with no spur at all. These latter pipes were popular in North America from about 1720 to 1820 and are believed to have been exported by Bristol pipemakers.

During the mid eighteenth century extra-long pipes became fashionable with the gentry. These were called *aldermen* but later became more generally known as *straws*. The stems were around 18 to 24 inches (450 to 600 mm) long and the bore diameters averaged $\frac{3}{32}$ inch (2.4 mm). As far as is known, these were the first pipes to have been given a specific name during their period of use.

Some London pipemakers were producing heraldic pipes bearing the royal arms, City arms, the arms of companies and the Prince of Wales's feathers during the first half of the eighteenth century. These pipes showed a very high standard of skill by the mouldmakers but designs tended to degenerate as the century wore on. By 1750 pipes were being made with masonic emblems on the bowl as well as with designs representing names of public houses and regiments. A leaf pattern on the seams of the bowl was common, but otherwise decorated pipes were rare. The aldermen, straws and some short plain pipes were the order of the day until well into the nineteenth century.

After 1850 the very long 'yard of clay' was introduced; as the name suggests, these pipes were about 36 inches (900 mm) long. They could not have been very practical and were little more than a passing phase. To assist the smoker, the stems were either marked with a label or twisted at the point of balance. Later in the century these long pipes were more popularly called *churchwardens,* although shorter versions (called *short church-wardens*) were eventually marketed. It has been said that the name 'churchwarden' was the invention of Charles Dickens. There is no proof of this, but he can perhaps be held responsible for perpetuating the name.

In the second half of the century the production of decorated pipes greatly increased and the Victorian businessmen were quick to exploit their use as an advertising medium. In addition to pipes illustrating events of the time and bearing all manner of slogans, there were designs depicting names of public houses, regimental badges, sporting activities, sailing ships, animals, fish, fruit, flowers and so on. Indeed there was such a host of different subjects that few customers would not have been able to purchase a pipe connecting them with their occupation or interests. These run-of-the-mill pipes were normally sold under the category of *fancy clays,* or *fancies.* The bore diameter of the stem for most pipes of this period was about $\frac{1}{16}$ inch (1.6mm) whereas the length of the stem and size of the bowl varied according to the design.

Most fancy clays had the design on the side of the bowl but there were some with the bowl shaped to represent the subject itself. Among these were designs touching on the naughty and macabre, such as the chamber-pot and the human skull. Happily there were also many pleasant subjects portrayed including heads of well-known comedians and jockeys and other famous characters of the day. Various versions of the heads of dragoons and negroes were popular. These two designs were illustrated in some of the old French catalogues (the dragon was referred to as *Cuirassier Anglais*) and were probably of French origin.

Another trend was brought about by competition from the beautifully carved meerschaum pipes with their comfortable amber mouthpieces. This forced the clay

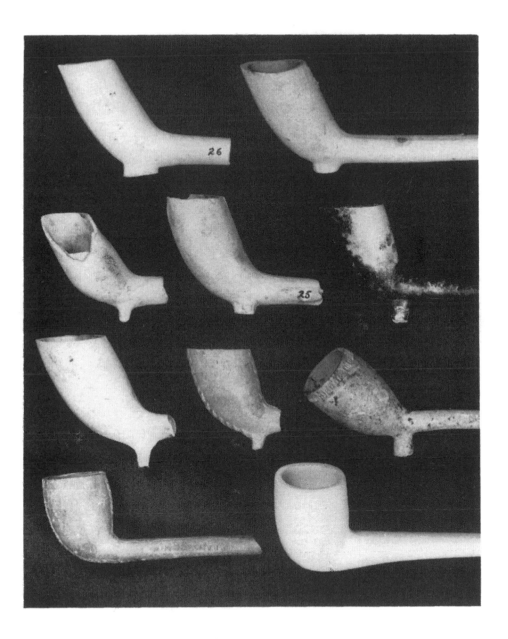

A selection of pipes dating from about 1690 to 1930 (top to bottom and left to right): c. 1690-1720, long parallel bowl with large pedestal spur; c. 1720-40, long bowl with top of bowl in line with axis of stem; c. 1720, Broseley-type bowl; c. 1720-50, bowl with thin wall section and shallow pedestal spur; c. 1800, bowl with long pedestal spur; c. 1840-60, large and very thin-section bowl with long flat-bottomed spur; c. 1840-70, small bowl with leaf pattern; c. 1860-90 Dutch-type bowl with heavy spur; c. 1880, cutty pipe; c. 1930, briar pipe.

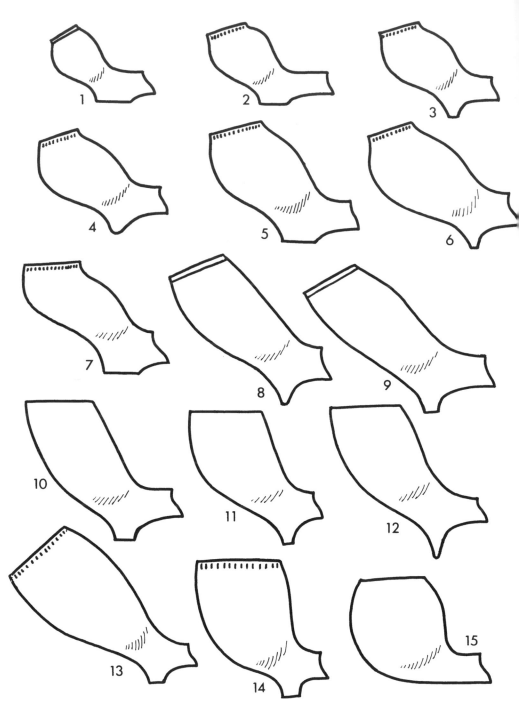

Bowl shapes: 1, c. 1580-1610, heart-shaped base; 2-3, c. 1610-40, flat bases and development of spurs, milling common; 4, c. 1640-60, small increase in size; 5-6, c. 1660-80, notable increase in size; 7, c. 1660-80, west country style; 8-9, c. 1680-1710, development of long bowls; 10, c. 1700-70, top of bowl parallel to stem; 11, c. 1770-1820, thin and brittle walls, pedestal spurs; 12, c. 1810-40, long pointed spurs; 13, c. 1850-1910, Dutch style, copied by some English makers; 14, c. 1850-1910, Irish style, made by some English makers from standard type mould; 15, c. 1860-1930, copy of briar.

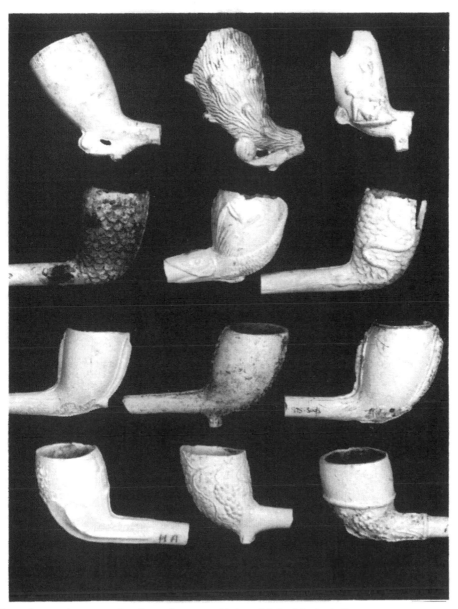

Late nineteenth-century pipes (top to bottom and left to right): plain football; rustic football; football scene; three angler's pipes; three cutty pipes with oak-leaf pattern; scalloped bowl with roses, thistles and shamrocks; two with grape vines.

pipemaker to introduce intricate designs of a similar nature: the eagle's claw clutching an egg (represented by the bowl) and a hand holding a wine glass are two well-known examples. They also produced the popular *character* or *portrait clays* of famous people, including members of the royal family, and many versions of the Victorian lady in her picture hat. These portrait clays were normally finished in a baked varnish (to produce the smoked meerschaum effect) and they were fitted with a vulcanite mouthpiece or else the clay stem was given a shapely bend after moulding. Towards the end of the nineteenth century imitations of the calabash and briar pipes were also made.

Despite the influx of these elaborate imitations, the nineteenth-century working man preferred his ordinary short clay, which was very cheap and often given away with a pint of beer by the local publican. The shorter pipe had the advantage of reducing the load on the teeth when smoking and working at the same time. This new habit (previously it was usual for a clay pipe to be smoked at leisure with the stem supported in the hand) brought about the production of special short pipes such as the Scottish *cutty* and the Irish *dudheen* although many a pipe was shortened by breaking off the unwanted portion of stem to suit individual needs. Before leaving the factory the ends of the stems were normally treated to prevent the lips sticking to the porous clay; to overcome the loss of this treatment when shortening the stem, the owner would have either dipped the broken end in tea or beer or carefully bound it with thread. Short clays were often referred to in the north of England as *nose warmers*.

The use of clay pipes for blowing bubbles was a popular pastime with children until about 1930. Pipes for this purpose were of the cheaper variety and were often to be found in penny packets of sherbet dab. Clay pipes were still being made for the bubble-blowing market until the 1950s and about that time John Pollock & Company introduced coloured pipes with the aim of restoring their popularity. Sadly, the much cheaper plastic imitation and the more recent idea of a plastic ring at the end of a stick finally prevailed. Clay pipes were also used as targets at fairground shooting galleries and were even used in party games. The long stem of a churchwarden sometimes came in handy as a drinking straw.

A popular sport at large public meetings, such as Derby Day, was to knock a clay pipe out of the mouth of a head made up from old rags. Commonly known as Aunt Sally, the head, presumably representing that of a woman, was perched on the end of a pole with a clay pipe sticking out of its mouth. The object of the sport was to knock the pipe out of the mouth with the use of a short, smooth round stick without damaging or hitting the head. The reason behind the sport is not known, but it could well have been intended as a deterrent for lady smokers. One much earlier function of the clay pipe was as an emergency powder measure for loading muskets during the Napoleonic wars. Pipemakers living close to naval ports may well have made pipes with the correct capacity in mind.

Two dodges favoured by navvies and other labourers were the use of a metal trouser button at the base of the bowl to act as a filter and the fitting of a metal bottle cap, pierced with holes, on top of the bowl to keep the rain out.

By 1914 clay-pipe manufacture as an industry had virtually come to end, leaving only a few well-established makers to meet the small but continuing demand. But there has always been a demand for the clay pipe from those who have enjoyed its cool smoke and slow burning properties and by masonic and other societies for their after-dinner smoke. They are still used by the Royal Antediluvian Order of Buffaloes during the initiation ceremony of new members.

During the 1960s there was a marked resurgence in the use of the clay pipe, caused by the formation of the Pipe Club of Great Britain. Several branches of the club were established up and down Britain and similar clubs were formed in France and in Australia. Competitions were held at branch meetings to find the smoker who could keep a 9 inch (225 mm) long clay pipe going the longest without relighting when filled with a regulated quantity of tobacco. Competitions were also held at national level and in consequence there was a good demand for the traditional clay pipe. The club was disbanded in 1978.

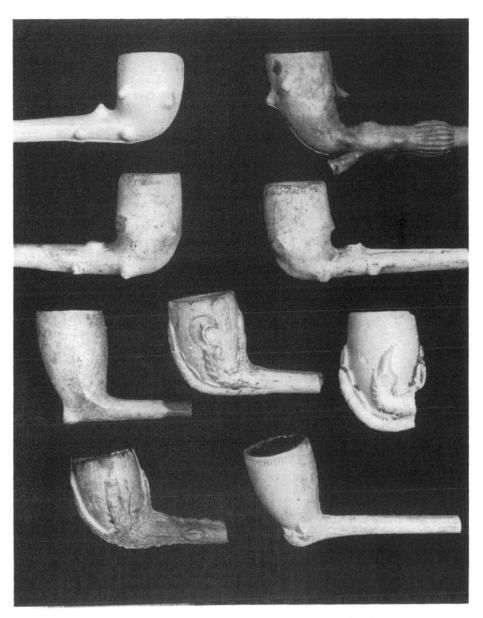

Late nineteenth-century pipes, showing some variations in the thorn and claw designs.

How Clay Tobacco Pipes Were Made

2. Forming the rough shapes

Balls of clay were rolled by hand and afterwards rolled out to form the rough shape of the pipe to be moulded. These were called 'rolls'.

3. Moulding.

A piercing rod was passed through the shank of a roll and placed in a two-piece mould as shown.

The assembled mould was then placed in a 'Gin Press' and the lever pulled down to form the hollow in the bowl. After this the piercing rod was withdrawn and the pipe removed from the mould.

4. Trimming and Firing.

When the pipes were dry any rough edges were removed with a trimming knife and the pipes placed in saggers ready for firing.

The saggers were then stacked in the kiln and fired to about 950° centigrade.

5. Finishing.

After firing the pipes were checked for flaws and the mouthpieces coated with wax or lacquer to prevent the smoker's lips sticking to the clay.

1. Preparing the Clay.

The clay was washed in wooden or copper tubs to remove dirt and stones and placed on boards to mature and dry.

When dry the clay was worked by beating with an iron bar to remove air.

A group of pipemakers outside Henry Leigh's pipe manufactory at Portchester, Hampshire, c. 1865.

PIPEMAKERS

Little is known about pipemakers until about 1619 when they were strong enough, in London at least, for James I to grant a charter of incorporation to the tobacco pipemakers of Westminster. The charter was renewed by Charles II in 1663. The charters protected the interests of pipemakers by governing the laws of trading within the city, controlling the training of apprentices and regulating the supply of clay.

Masters of the craft jealously guarded the secrets of their trade and ensured their apprentices did likewise. There was always the possibility of unskilled workers branching out on their own, particularly other tradesmen such as bakers and alehouse keepers. The Company of

Pipemakers, soon after its incorporation, had the right to enter dwelling-houses to break up unlawful pipemaking. This was deemed necessary not only to protect members but also to prevent the sale of inferior products.

Early pipemakers were beset by many problems, such as the transportation of clay, the availability of fuel for firing and the distribution of the finished pipes. Their troubles were increased by the innkeepers' practice of cleaning pipes after use by their customers — the pipes were gathered together in a small iron frame made for the purpose and placed in the hot embers of a fire. They may have been taken to the forge of the local smithy, but care would have been necessary lest the steam from residual moisture and oils of the tobacco caused the pipes to shatter. Records show that most seventeenth-century pipemakers were poor and it was not unusual for a pipemaker to travel from town to town to escape unhealthy competition. Several London makers migrated to the eastern ports, where there was a good coastal trade.

By 1650 there were at least a thousand pipemakers in London alone and many others operating in other towns such as Bristol, Broseley, Chester, Gateshead, York and Hull. The industry had also spread to other parts of the British Isles and to Holland; early Dutch pipemakers had English names and are thought to have been among the Puritans who settled in Holland at the beginning of the century seeking freedom of religion.

Between about 1670 and 1740 the popularity of snuff-taking caused a recession in the trade. This new habit was brought over from France by the followers of Charles II on his return from exile in 1660; the courtiers there considered it to be more elegant to sniff powdered tobacco than to smoke it. Otherwise the trade continued to expand so that most towns in England had at least one pipemaking family. In the nineteenth century factory production was introduced and led to the decline of the trade as a cottage industry. Nevertheless, throughout the history of making clay tobacco pipes the master craftsman would have kept his business on a family footing and it was not uncommon for his widow or daughter to continue trading after his death.

Some pipemakers augmented their earnings by producing other objects in clay, such as hair-curlers during the seventeenth and eighteenth centuries. More recently blocks of hearth-stone were prepared for the proud housewife to keep her front doorstep white and clean. A good example is that of Henry Leigh & Company of Portchester, pipemakers from 1840 to 1932, who were also whiting and putty manufacturers as well as wholesale dealers in bath-brick and hearth-stone. There is also evidence of pipemakers running a second trade in addition to making pipes, but this was usually the sale of ale or tobacco.

There were only a few pipemaking concerns left during the early years of the twentieth century and these slowly disappeared leaving but one survivor, John Pollock & Company of Manchester. Founded in 1879, the company finally closed down in 1992 on the retirement of the proprietor, Gordon Pollock. The last of the London makers was probably Charles Crop & Sons of Brooksby Walk, Homerton. Founded in 1856, the firm ceased production in 1924 and was responsible for the production of many fine portrait clays. The Southorn family of pipemakers in Broseley, Shropshire, closed down their factory in the 1960s. The factory still stands, as well as the bottle kiln, and there are plans to reopen the buildings as a working museum.

The pipemaking industry in Scotland was well established by the beginning of the eighteenth century. The earliest recorded maker was Stephen Bell of Edinburgh in 1649 and the last was William Christie of Leith in 1962.

J. Luther of Youghal, about 1687, is the only seventeenth-century Irish maker found so far. There are no known eighteenth-century makers, but at least twenty-three were operating from 1819 to 1917.

Very little is known about the industry in Wales but there were at least six makers between 1812 and about 1850. That there were earlier Welsh makers is suggested by the charter of 1663, which was addressed to the 'Tobacco Pipemakers in the Cities of London and Westminster and the Kingdom of England and the Dominion of Wales'.

Jersey and Guernsey each had at least one maker, and both of them are recorded as working in 1852. They were John Welsh of St Helier and W. S. Chaple of Bouet.

The recent revival in the use of clay pipes prompted the formation of the Pilgrim Pipe Company in 1972. Operating from Skegness, the company mass-produced 9 inch (225 mm) long pipes and a few character clays using moulds which once belonged to William Christie of Leith. It was not long, however, before the supply became greater than the demand and the company closed in 1975.

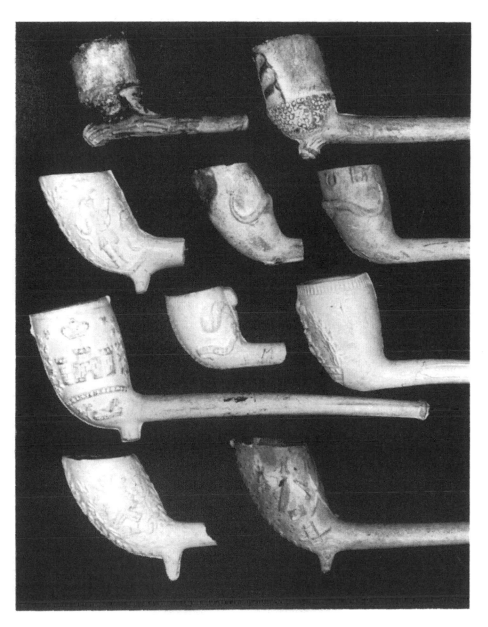

Nineteenth-century pipes (top to bottom and left to right): two acorns; man with bow and arrow; two RAOB pipes with buffalo horns; military crest (The Inniskilling Regiment); Prince of Wales's feathers; military crest; two masonic pipes.

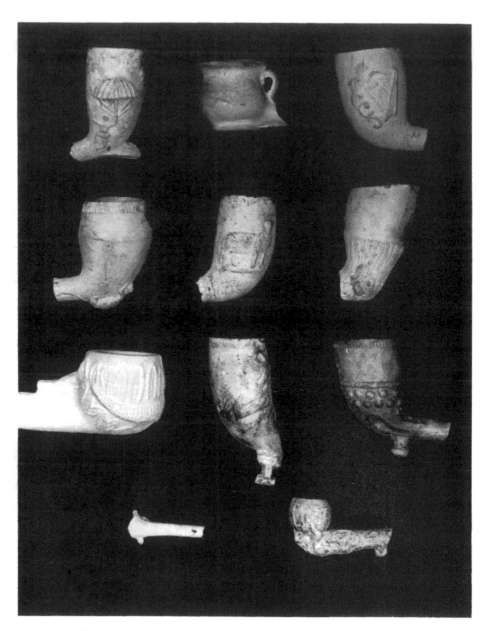

Late nineteenth-century pipes (top to bottom and left to right): parachuting; chamber pot; Irish harp; rats on bowl and stem; a sheep; pony's hoof; crocodile; Queen Victoria with a crown-shaped spur; scallops and dots; part of a miniature; a cigarette pipe depicting Atlas supporting the universe.

16

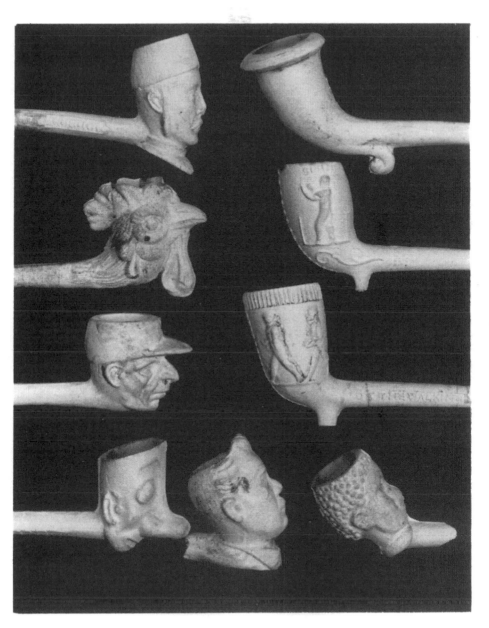

Late nineteenth-century pipes (top to bottom and left to right): General Gordon; a horn; a cockerel; boxers (Sullivan and Smith); a jockey (probably Fred Archer); walkers; Ally Sloper (cartoon character); Joseph Chamberlain (statesman); head of a negress facing the smoker.

Late nineteenth- and early twentieth-century pipes (top to bottom and left to right): proud lady; Mrs Sarah Wilson (poetess), an expensive portrait clay produced from a three-piece mould and finished in baked varnish; a lady motorist; King Edward VII, another portrait clay from a three-piece mould; William Gladstone (statesman); Ally Sloper.

A pipemaker preparing a mould for the gin press at Henry Leigh's factory at Portchester, Hampshire. Note the dozening board on the right of the bench.

PIPEMAKING

Except for the earliest clay pipes, which were made by hand, the basic method of manufacture has remained the same since the introduction of the two-piece mould shortly before 1600. The first moulds were made of brass (though it is thought that some were made of wood) but iron moulds soon followed and were in general use by 1750. Unfortunately it is not known if any moulds exist which were made before the end of the eighteenth century, but several nineteenth-century moulds can be seen in museums.

The making of moulds was a specialised art, and, except for a few registered designs, most nineteenth-century moulds were available to the pipemaker by illustrated catalogue. Because metal moulds had a long life many old styles overlapped the newer ones. However, because clay is abrasive, moulds eventually suffered from wear so that the bowls and stems became noticeably larger and the details in the design less defined.

Early pipemakers in London and other large towns would have used clay supplied by the quarries in Cornwall, Devon and Dorset, but some were lucky enough to have lived near workable local deposits. Two well-known local deposits were at Broseley in Salop and Amesbury in Wiltshire. The latter was worked by the Gauntlet family for many years during the seventeenth century and their pipes were famous for their fine quality.

The pipes, several thousand at a time, were baked in an up-draught kiln made with bricks of fire-clay. The kiln was shaped like a cylinder with a domed roof and chimney, and it looked very like the bottle-kiln used by the potteries, but on a smaller scale. Originally charcoal or wood was used for fuel but the kilns were later modified to burn coal or coke. Because of

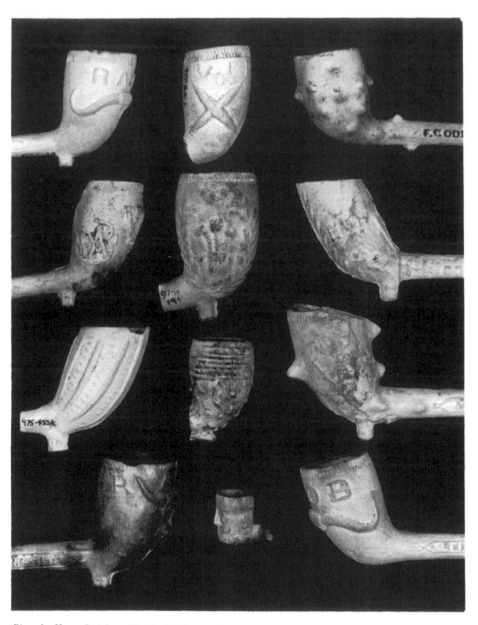

Pipes by Henry Leigh and Frederick Goodall of Portchester (top to bottom and left to right): RAOB pipe; HMS Vernon (Portsmouth submarine base), showing crossed torpedoes; a thorn pipe; an unusual RAOB pipe with the insignia on the side of the bowl; Prince of Wales's feathers; masonic pipe; ribs and dots; bands and dots; Mother Shipton (sixteenth-century witch), probably made for the public house of that name in Portsmouth; RAOB pipe with horns and head of buffalo; cigarette pipe depicting a bearded man; RAOB pipe.

Pipe mould from Henry Leigh's factory at Portchester.

the high cost of firing it was not unusual for two or three pipemakers to share one kiln.

Nineteenth-century pipes were loaded in containers called saggars and these in turn were stacked in the kiln. Introduced during the first half of the nineteenth century, saggars were either rectangular or circular and made from fire-clay. It is thought that previously home-made containers were used made up from old pipe stems (wasters) embedded in clay. The long-stemmed pipes had to be loaded in a vertical position with the bowls lowermost. The stems were supported by a pillar in the centre of the saggar and secured by a ring of soft unfired clay. After the first layer of pipes was positioned further layers were added and the wall of the first saggar raised as necessary by bottomless saggars. Pipes might also have been stacked directly on shelves or similar supports.

Except for a bench-vice or gin-press, only a few simple hand tools were required for the trade. These were piercing rods for making the bore in the stem, knives with specially shaped blades for trimming the mould flashes from stem and bowl, hand-stoppers for forming the hole in the bowl and various smoothing tools.

The gin-press was a vice with an overhead lever to which was attached a stopper to suit the bowl of the pipe to be moulded. With the mould held in the vice by either a screw or a 'flying handle', the lever was pulled down to allow the stopper to enter the bowl end of the mould. The stopper was attached to the lever by a bolt permitting fore and aft movement so as to allow for the circular movement of the lever.

21

A gin-press from Henry Leigh's factory. This was the only form of mechanisation used in pipemaking.

This simple but highly efficient apparatus was, as far as is known, the only form of mechanisation used in the moulding of clay pipes. The only limitation was its inability to form bowls that were set at an acute angle to the stem. This may have been why pipes were made with more upright bowls from about 1700, when the gin-press is thought to have been introduced.

Moulds used in conjunction with the gin-press would have had to be modified by providing an extension above the bowl end to act as a guide for the stopper. This addition can be clearly seen on nineteenth-century moulds, as can the cleft for the insertion of the knife when trimming the top of the bowl. The gin-press was probably in common use by 1750 but long after this there were still many pipes being made with the bench-vice and hand-stopper.

The first and most important stage in the manufacture of clay pipes is the preparation of the clay. Clay was received from the quarries in large lumps which were broken down into smaller pieces and washed in a large wooden or copper tub. After cleaning and the removal of stones and other foreign matter, the water was drained off and the clay placed on boards to dry and mature. When ready the clay was beaten with a heavy iron bar to expel any air and finally kneaded to a uniform mass. The next stage was to break the prepared clay into small blocks from which portions of clay were taken and rolled into rough shapes of the pipes to be moulded. These rough shapes, called rolls, were then placed in groups of twelve on boards called dozening boards and passed to the moulder. Women and children between twelve and fifteen years old performed this task and were known as

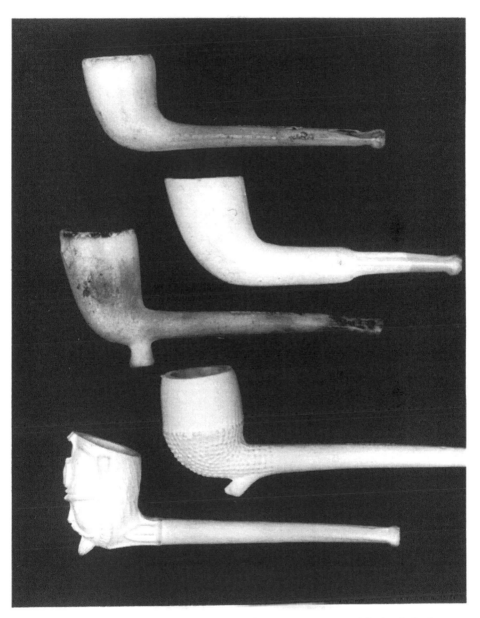

Pipes by Southorns of Broseley, showing three short plain pipes, an acorn and the head of a dragoon.

rollers.

After a short drying period, the moulder laid a roll on the bench and pierced the stem part with a brass or steel rod before placing the roll into one half of the mould. Both the rod and the mould were lightly greased to allow the finished pipe to be easily removed. The other half of the mould was positioned and the whole assembly was pressed in either the bench-vice or the gin-press. After forming the bowl with the hand-stopper, or pulling down the lever of the gin-press, the moulded assembly was then removed from the vice and the rod was manipulated to marry with the interior of the bowl which had previously been hollowed out by the action of the stopper. The moulder would then trim the excess clay left by the joins in the mould and from the top of the bowl before withdrawing the piercing rod and placing the pipe on a tray to dry.

When full of pipes, the trays were placed on drying racks where air could freely circulate. During the drying a wire was passed through the bore of the stem to prevent distortion of the clay when shrinking. In the case of long pipes requiring curved stems, the wires were removed before the pipes were completely dry and the stems were gently bent to the required shape. The pipes were afterwards passed to the trimmers and finishers who took off any rough edges with a sharp knife and polished the more expensive pipes with a wooden burnishing tool. The pipes were then inspected for flaws and placed in a saggar for firing.

The sequence for raising the temperature to about 900C (1650 F) and subsequent cooling took about three days to complete. The temperature had to be raised very slowly during the initial stages of firing, in order to drive off any residue moisture in the clay before it reached boiling point, otherwise the clay would explode. When the pipemaker felt that the correct temperature had been reached he would remove a loose brick from the wall of the kiln and take out a piece of clay to check its temper.

After firing, the last stage of manufacture was to treat the mouthpiece of the stem so that the smoker's lips would not stick to the porous clay. The cheapest of pipes, if treated at all, were dipped in water containing a little pipe-clay in solution and then given a polish. Better quality pipes were coated with a mixture of soap, wax and gum. It is doubtful if pipe stems were treated as a normal part of the process much before the nineteenth century, but some eighteenth-century pipes have been found with the stems dipped in glaze. During the last quarter of the nineteenth century it was common practice to treat the stem with a red sealing wax by heating the end of the stem on a hot plate (heated by a gas flame) and then rolling it on to the wax. Latter-day pipes are usually treated by applying transparent or coloured lacquer with a brush.

Finally the pipes were packed in wooden boxes, using wood shavings or sawdust for protection, and dispatched to the customer.

Pipe wax from the workshop of Richard Norwood of Eton c. 1914.

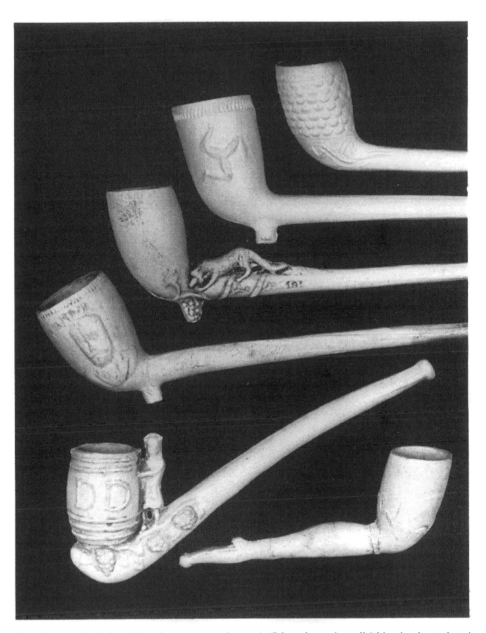

Pipes by John Pollock of Manchester (top to bottom): fish scales and scroll (this pipe is marketed under the name of 'social pipe', which may have been given to most pipes with similar nondescript patterns); Isle of Man; fox and grapes; Charles Parnell of Ireland; Dirty Dick; a lady's leg.

Three views of an eighteenth-century pipe made in Gouda by Willem Noppen commemorating the birth of Prince William V of Orange, 8th March 1748.

EUROPEAN PIPES AND PIPEMAKERS

The popularity of the English clay pipe soon spread to the Low Countries, Germany and, to a lesser extent, Denmark and Sweden. In France, however, clay tobacco pipes do not appear to have been widely used until well into the eighteenth century, probably because the French preferred snuff-taking, but French pipemakers then began producing what were known at the time as 'French clays' and are now generally referred to, along with the character and portrait pipes, as 'figurals'.

The first known pipemakers in Holland were Englishmen who had left their country when James I discouraged the use of tobacco and enforced restrictions in the manufacture of clay tobacco pipes outside the City of London. Also it is quite probable they were joined in the trade by fellow countrymen who had settled in Holland seeking religious freedom, as well as those who had served in the armies of Prince Maurice.

The earliest mention of pipemakers in Holland was of two Englishmen in Amsterdam, William Jorreson Boyesman and Thomas Lourens, in a deed of 1611. Shortly after that date there is evidence of Dutch makers working alongside English makers in Amsterdam and other Dutch towns, notably Haarlem, Rotterdam, Groningen, Leiden and Gouda. The earliest known pipemaker in Gouda, which became the major centre of the new industry, was

also an Englishman, who set up his trade in 1617. Known by the Dutch as Willem Barends, his original name was thought to be William Bearnelts. These early pipemakers from England were further encouraged to ply their trade by the fact that, in 1627, another Englishman, Franchoys Jorisz Davids, who was a surgeon with a practice in Rotterdam, received a patent for the sole supply of pipe clay in Holland.

Fierce competition and squabbles soon developed between manufacturers of both nationalities, particularly over the copying by unscrupulous makers of the marks of other makers of better quality – a practice not unknown in England. To overcome this, pipemakers were required to be registered by the city authorities, each maker having an approved mark or motif with which to identify his product. Records of registrations and marks have been well preserved and documented in Holland and so collectors of Dutch pipes can readily identify pipes with their makers.

Almost from the beginning the Dutch adopted a style of their own (see figure 13 on page 8) and, being more conservative than the English, maintained this basic shape until well into the nineteenth century. As with English pipes, the general rule is that the smaller the bowl and the bulkier the stem the earlier the pipe is. By about 1850 there were a number of large manufacturing companies, which were later forced to compete for a share of the market with the increasingly popular French clays.

One of the best-known pipemakers in Gouda was the family of Goedewaagen, whose workshops did not close until the late twentieth century. Another well-known family of pipemakers from Gouda is that of P. J. van der Want. Their successors, MessrsWestraven, manufacturers of tiles and other earthenware products since 1661, continue to make traditional pipes as well as clay pipes of a more modern concept, using the slipware process.

By about the middle of the seventeenth century several Dutch and some English pipemakers had settled in what is now Belgium, probably to escape the stringent rules of the registration system in Holland. Otherwise little is known of pipemaking in this region until the late eighteenth century. Among the best-known pipemakers were the de Bevere family, who made pipes in Courtrai from 1772 until 1950. Three others worth noting are Levèque of Andenne, from 1830 to 1935, the Knoedgen family of Liège, from 1770 to about 1857, and Leonard of Andenelle, who began making pipes in 1930. A variety of pipes is still being made at the Leonard workshops, including the Jacob pipe mentioned below.

Not much is known about the early clay-pipe industry in Germany, except that many Dutch-styled pipes have been found during archaeological excavations. These were either imported, made by Dutch settlers or copied by German pipemakers. There were a number of thriving pipemaking concerns in Germany during the middle of the eighteenth century producing highly decorated, traditionally styled pipes as well as figurals. About this time the purely German porcelain or Tyrolean pipe, which became immensely popular with German and Austrian pipe smokers, was being produced. It consisted of a glazed earthenware bowl and reservoir fitted with a wooden stem and a horn mouthpiece. The bowls on these pipes were often elaborately decorated with hand-painted pictures of hunting and other scenes.

As already mentioned, French pipemakers were famous for their 'French clays', of which there was an almost infinite variety of designs depicting famous and fictional characters from all walks of life, as well as animals, flowers, transport and a host of other subjects. The majority of these pipes consisted of a moulded bowl-head with a detachable wooden stem and horn mouthpiece. French pipes were also noted for the colouring of details with enamel glaze, with white enamelled eyes and black dots for the eyeballs.

Two major pipemaking concerns in France were those of Fiolet of St Omer, from 1765 to 1921, and Gambier of Givet, from 1780 to 1926. The firm of Gambier was the larger of the two, producing several million pipes a year from well over two thousand different designs, many of which were made from three-piece moulds. In 1848 Gambier had two factories employing around 330 people, and by 1867 there were as many as 600 employees. Another equally well-known, but less long-lasting, maker in St Omer was that of Duméril, Leurs (1845-85).

The most popular of all the French clays was the pipe depicting the bearded biblical patriarch Jacob with a turban-like head-dress. French and Belgian pipemakers produced millions of the 'Jacob' pipes alone and, because of fierce competition, several variations of the subject were made in addition to differences in the inscription embossed across the hatband. The more common of those used were: JE SUIS LE VRAI JACOB ('I am the real Jacob') and DEMANDEZ JACOB ('Ask for Jacob'). There was also a similar model, produced in smaller quantities, bearing the name VERITABLE ABRAHAM.

Late nineteenth-century French clay with the head of a greyhound by Duméril, Leurs of St Omer.

DATING PIPES

Work recently carried out by archaeologists has made it possible to date English clay pipes accurately to within twenty years or so. This has been achieved by a close study of bowl shapes in conjunction with maker's marks, the relationship of pipes and other objects found with them during excavations and documentary evidence.

The drawing on page 8 shows some typical bowl shapes from about 1580 to 1930. When establishing the date of new finds it is helpful first to group them in order of stem-bore size and thickness (the larger the bore and the thicker the stem, the earlier the pipe is likely to be) and then make a further assessment by checking the size and shape of the bowls.

Maker's marks can sometimes establish a closer date, but care has to be taken when only the maker's initials appear since there may have been others using the same initials for a different period. The normal method of marking pipes at the beginning of the seventeenth century was for the maker to stamp his initials or his complete name on the heel of the bowl using a signet ring or stamp. In most cases the initials were enclosed in a circle or heart-shaped border and usually incorporated a decorative scroll-like design. A number of pipes have also been found bearing the maker's personal symbol such as the gloved hand used by the Gauntlet family of pipemakers in Wiltshire.

Later in the seventeenth century cartouche marks were stamped on the side or rear of the bowl. These were in the form of circles enclosing the maker's initials or name and also the name of the town. On rare occasions the date was included. About this time pipes were also being made with the maker's initials incorporated in the mould on each side of the spur. This method of marking became the normal practice until about 1860, when most marked pipes showed the maker's name on the side of the stem and sometimes his address or the name of the town in which he worked. Some makers, however, continued to use the cartouche mark, or the initials on the spur, until the end of the nineteenth century.

The initial of the maker's first name was usually shown on the left-hand side of the spur (when looking at the pipe with the bowl facing to the left) and the initial of his second name on the right-hand side. The letter I was invariably used in place of the letter J. Some pipes have been found showing initials superimposed on earlier ones indicating the continuing use of moulds by the pipemaker's son or apprentice.

In examining the names on the stems of nineteenth-century pipes, caution should be taken with double names and those ending in '& Co' or '& Son' as they could be the names of tobacconists or wholesalers. This can sometimes be verified from local trade directories.

As mentioned earlier, seventeenth-century Dutch pipes are more readily dated by the registration marks of the makers. Another useful, if more remote, means of dating Dutch pipes in particular is by reference to pipes shown on prints of seventeenth-century Dutch and Flemish painters, notably those by Adrian Brouwer (1605-38), Jan Steen (1629-79) and David Teniers (1610-90). Some caution has to be exercised, however, when dating pipes shown in pictures by nineteenth-century English artists, especially those depicting eighteenth-century public-house scenes, which invariably show contemporary styles. Also, many seventeenth-century English engravings portray, incorrectly, a stylised bowl shape looking very much like an egg-cup perched vertically on the end of a stem.

Eighteenth-century Dutch pipes showing the makers' registration marks.

28

COLLECTING PIPES

The remains of clay tobacco pipes can be found in almost any part of Britain, for an enormous quantity of pipes has been made over the past three hundred years or more; one nineteenth-century factory alone would have produced several thousand every day of the week. Even so, there is a greater likelihood of finding pipes or their remains in places which have been continuously occupied, particularly in old industrial areas.

Like the bottle collector, the clay-pipe hunter should look out for any known Victorian dumping ground, which might well yield bowls and sometimes complete pipes. Other places to look are tidal rivers and estuaries. Because of the difficulty in establishing the origin of finds from large Victorian dumps, the more rewarding finds, for the historian at least, are those from sites where the pipes were last used. Garden beds, ploughed fields, hedgerows, streams and some small local dumps are good examples. A river bed adjacent to a bridge is always worth a look, especially if there is a public house close by — many an imbiber of the local brew would have cast his old 'clay' over the side on his way home.

A great many seventeenth-century pipes have been found in fields where the troops of Charles I or the Parliamentarians were quartered during the Civil War, so any known battle site or marching route is a likely area. Smaller numbers of pipes of the eighteenth or nineteenth centuries could possibly mark the site of an old fairground or gypsy encampment. The occasional piece of stem found in a field is probably the remains of the ploughman's pipe.

Those living in old towns or villages should look out for any excavations during road repairs or the erection of new buildings. Complete pipes are often found under the floorboards or in the rafters of old buildings, where they were left during the original construction or renovations. Before looking for pipes on private property, however, one must always obtain the owner's permission.

Owing to the light density of clay pipes, they are often found lying on the surface, so there is rarely any need to dig for them: it is only necessary to wait for the garden bed to be dug or the field to be ploughed, ideally after a good heavy shower of rain, which will wash the dirt from any remains of clay pipes, making them clearly visible.

There are always many more pieces of stem to be found than bowls and there are two possible reasons for this: the ends of the stems were often broken off during the useful life of the pipes, and the remaining piece of pipe, with the bowl attached, would have been finally discarded in a fire or rubbish bin.

From time to time clay pipes found in Victorian rubbish dumps are unearthed looking as if they were new. This is usually because the heat from fires when the rubbish was burnt, though sufficient to melt the glass of old bottles, did no harm to clay pipes (or any other ceramic products originally baked at much higher temperatures) other than to burn off any discolouration or other signs of use.

The revived Victorian pipes produced today, as well as a number of reproductions, all have a valued place in a collection so long as they are properly identified. They should preferably be looked for in tobacconists or gift shops where normal retail prices are charged and where the shopkeeper may be able to give some information on the maker.

Although some good finds can be purchased from second-hand and antique shops, it is better to look there only after some experience on the subject has been gained. Some revived pipes have been given a special ancient look before leaving the maker's workshop: distinguishing between these and the genuine article requires a practised eye.

It is most important to identify finds temporarily until they can be cleaned and dated. After cleaning they can be individually marked with black ink or grouped in a labelled box or bag. Dirt and mud is easily removed from bowls and stems with warm soapy water and a soft brush. It may be necessary to remove dirt from the bore of the stems with a thin piece of wire. Rust marks and other discolourations are best left since they are evidence of the pipe's age and history. Rust marks are usually caused by the submersion of the pipe in a river bed for several years, and a uniform dirty grey to black appearance can be caused by the tobacco oils absorbed by the clay over a long period of use. Unfortunately evidence of this nature is almost always removed from old pipes by the action of the soil and the weather.

A few Victorian curios are still to be found, such as the coiled pipe and the multi-bowled pipe. Many of these were the work of apprentices for test-pieces and they can command high prices. Other unusual specimens are pipes with tiny bowls for smoking cigarettes and miniature pipes which could be of use only in dolls' houses. There are also pipes with oversize bowls known as 'cadger' pipes. These were used in fun when a free fill of tobacco was offered by an unsuspecting friend.

For those who want to repair broken pieces of pipe a number of modern clear adhesives will do, but it is sometimes best to leave the pieces as found, either grouped together on display or kept in a box or tray.

The best way to display clay pipes depends largely on space available. Wall cabinets are perhaps the most practicable since they take up the least space and the pipes can be attractively arranged by securing them to the back and sides of the cabinet with terry clips or wire. On the other hand the simplest method is to use a table-top cabinet, which can easily be made from picture-frame mouldings and a hardboard base.

FURTHER READING

Since this book was first published there has been a continuing increase in books and other forms of literature on the subject of clay tobacco pipes. Most of these can be obtained from public libraries and museums. Some of the better known publications are listed below:

BAR 239. *The Archaeology of the Clay Tobacco Pipe.* 1994.
BAR 246. *The Development of the Clay Tobacco Pipe Kiln in the British Isles.* 1996.
Brongers, Georg A. *Nicotiana Tabacum.* Groningen, 1964.
Duco, Don. *Genealogie Goedwaagen.* Amsterdam, 1977.
Dunhill, Alfred. *The Pipe Book.* 1969.
Dunhill, Alfred. *The Gentle Art of Smoking.* Max Reinhardt, 1969.
Scott, Amoret and Christopher. *Smoking Antiques.* Shire Publications, reprinted 1996.
Walker, Iain C. *The Bristol Clay Tobacco-Pipe Industry.* Bristol Museum, 1971.

Note: BARs (British Archaeology Reports) can be obtained from Hadrian Books, 122 Banbury Road, Oxford OX2 7BP.

THE INTERNET

There are now a number of interesting websites on the internet giving information on clay tobacco pipes, by collectors, organisations and other enthusiasts. These can easily be found by typing in 'clay tobacco pipes' in the search column. The sites include details of the Society for Clay Pipe Research (SPCR), with membership enquiries to Peter Hammond at claypipepeter@aol.com. More information can be found at: www.dawnmist.demon.co.uk/scpr.htm

This website also illustrates and describes the clay pipes made in the workshop of Heather Coleman, co-ordinating information about clay pipes with experts and enthusiasts all over the world.

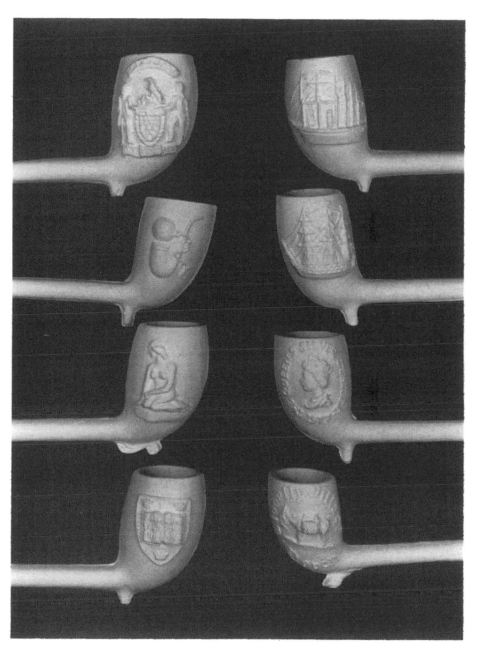

Designs by Eric G. Ayto (top to bottom and left to right): the Cornish Arms; SS Great Britain; Pipe Club of Australia; HMS Victory; the Little Mermaid; HM Queen Elizabeth's Silver Jubilee; Oxford University; Clovelly village.

PLACES TO VISIT

Abbey House Museum, Abbey Road, Kirkstall, Leeds LS5 3EH.
  Telephone: 0113 230 5492. Website: www.leeds.gov.uk/abbeyhouse
Amberley Museum, Houghton Bridge, Amberley, Arundel, West Sussex BN18 9LT.
  Telephone: 01798 831370. Website: www.amberleymuseum.co.uk
Bewdley Museum, The Shambles, Load Street, Bewdley, Worcestershire DY12 2AE.
  Telephone: 01299 403573.
Bristol City Museum and Art Gallery, Queen's Road, Bristol BS8 1RL.
  Telephone: 0117 922 3571. Website: www.bristol-city.gov.uk/museums
The British Museum, Great Russell Street, London WC1B 3DG.
  Telephone: 020 7323 8299. Website: www.britishmuseum.org
Broseley Pipeworks and Clay Pipe Museum, Duke Street, Broseley, Shropshire TF8
  7AW. Telephone: 01952 433522. Website: www.ironbridge.org.uk
Curtis Museum and Allen Gallery, High Street, Alton, Hampshire GU34 1BA.
  Telephone: 0845 603 5635. Website: www.hants.gov.uk/museum/curtis
Gosport Museum, Walpole Road, Gosport, Hampshire PO12 1NS.
  Telephone: 023 9258 8035. Website: www.hants.gov.uk/museum/gosport
Guernsey Museum and Art Gallery, Candie Gardens, St Peter Port, Guernsey GY1 1UG.
  Telephone: 01481 726518. Website: www.museum.guernsey.net
Guildford Museum, Castle Arch, Guildford, Surrey GU1 3SX.
  Telephone: 01483 444750. Website: www.guildford.gov.uk/museum
Museum of Edinburgh, Huntly House, 142 Canongate, Edinburgh EH8 8DD.
  Telephone: 0131 529 4143.
Museum of London, 150 London Wall, London EC2Y 5HN.
  Telephone: 020 7001 9844. Website: www.museumoflondon.org.uk
Norris Museum, The Broadway, St Ives, Cambridgeshire PE27 5BX.
  Telephone: 01480 497314. Website: www.norrismuseum.org.uk
Oxfordshire Museum, Fletcher's House, Park Street, Woodstock, Oxfordshire OX20
  1SN. Telephone: 01993 811456. Website: www.tomocc.org.uk
Priest's House Museum, 23-27 High Street, Wimbourne Minster, Dorset BH21 1HR.
  Telephone: 01202 882533. Website: www.priest-house.co.uk
Salisbury and South Wiltshire Museum, The King's House, 65 The Close, Salisbury,
  Wiltshire SP1 2EN. Telephone: 01722 332151.
Trowbridge Museum, The Shires, Court Street, Trowbridge, Wiltshire, BA14 8AT.
  Telephone: 01225 751339.
Winchester City Museum, The Square, Winchester, Hampshire SO23 9ES.
  Telephone: 01962 848269. Website: www.winchester.gov.uk/heritage

The following European museums are known to have permanent displays of clay pipes and other tobacco-related artefacts:
Holstebro Museum, Sonderbrogade 2, DK 7500 Holstebro, Jutland, Denmark.
Musée Communal de la Ceramique d'Andenne, 29 rue Charles Lapierre, B-52200 Andenne, Belgium.
Musée Communal de la Vie Wallone, Cours des Mineurs, Liege, Belgium.
Musée de Sandelin, St Omer, France.
Musee-Galerie de la Seita, 12 rue Surcouf, 75007, Paris, France.
Museum De Moriaan, Westhaven 29, NL-2801 PJ Gouda, Holland.
Niemeyer Tabaksmuseum, Brugstraat 24-26, NL-9711 HZ Groningen, Holland.
Pijpenkabinet, Oude Vest 159a, NL-2312 XW Leiden, Holland
Stedelijk Museum voor Pijp en Tabak, Marktstraat 100, B-8730 Harelbeke, Belgium.
Tabaks Museum, Brikkenmolen, Koestraat 63, Wervik, Belgium.
Tobacco Museum, Bunde, Germany.
Tobacco Museum, Gubbhyllan, Skansen, Stockholm, Sweden.

Printed and bound by CPI Group (UK) Ltd, Croydon, CR0 4YY

16/10/2024

01774983-0001